Mobile Repair Manual
Understanding the Scope of Mobile Phone Repair

© 2022 Felix Raluchukwu Okafor All rights reserved. No part of this book may be reproduced, stored, or transmitted in any form or by any means without prior written permission from the author, except for brief quotations in reviews or articles.

Dedication

This book is dedicated to my mentor, guide, and the person who ignited my passion for mobile phone repair—my esteemed boss, Igboke Chiwendu Samuel, affectionately known as Oga Sammy.

Your unwavering commitment to teaching and nurturing my skills in the field of mobile phone repair has been the cornerstone of my journey. Your patience, expertise, and leadership have shaped me into the technician I am today.

I am forever grateful for the knowledge you imparted, the challenges you posed, and the opportunities you provided. Your mentorship has not only enriched my professional life but has also instilled in me a deep love for this craft.

As I continue to grow in this field and embark on my entrepreneurial path, I carry your wisdom and teachings with me. This book is a tribute to your invaluable guidance and serves as a testament to the impact you've had on my life and career.

Thank you for being a mentor, a leader, and an inspiration. Your legacy lives on through the knowledge and skills you've passed on to me.

With profound gratitude,

Felix Raluchukwu Okafor

Acknowledgments

I would like to extend my heartfelt gratitude to several individuals who have played pivotal roles in my journey as a mobile phone repair technician.

First and foremost, I dedicate this acknowledgment to my beloved mother, Beatrice Nnoye Okafor. Though she is no longer with us, her unwavering support and encouragement continue to inspire me every day. May her soul rest in eternal peace.

I would like to express my appreciation to all the wonderful customers and patrons who have entrusted me with their mobile devices over the years. Your trust and support have been instrumental in my growth as a technician.

I also extend my gratitude to Mr. Shedrack, the owner of the current space where I operate my workshop. Your generosity and willingness to provide me with a place to pursue my craft are deeply appreciated.

To my dedicated team members, Chidubem and my brother Ifeanyi, your commitment to this field and your contributions to our success fill me with pride. Together, we make a formidable team.

Last but not least, I am thankful to all those who have believed in me, mentored me, and shared their knowledge and expertise. Your guidance has been invaluable.

To each and every one of you, I extend my sincere appreciation for your support, encouragement, and belief in my abilities. You have played an integral part in shaping my journey as a mobile phone repair technician.

With gratitude,

Felix Raluchukwu Okafor

Table of content

Understanding the Scope of Mobile Phone Repair 10

Limitations and Challenges 12

Importance of Mobile Phone Repair Technicians 13

Course Overview 14

Mobile Phones: An Overview 15

Standards and Regulations 16

Evolution of Mobile Phone Technology 17

Necessity in Modern Life 20

Safety Precautions in Mobile Phone Repair.............................. 22

Dismantling a Mobile Phone... 29

General Dismantling Procedure.. 33

Reassembling a Mobile Phone.. 37

Mobile Phone Components 44

Using a Multimeter for Fault Finding in Mobile Phone Repair .. 55

Common Mobile Phone Faults and Solutions........................... 60

Water Damage and Solutions for Mobile Phones..................... 68

Professional Solution for Water-Damaged Mobile Phones (Technician's Guide) 74

Advanced Troubleshooting Techniques in Mobile Phone Repair 79

Additional Resources for Mobile Phone Repair Technicians 85

Introduction to Mobile Phone Repair

Mobile phones have become an integral part of our daily lives, serving as communication devices, entertainment hubs, and tools for countless tasks. With the ever-increasing reliance on these devices, the need for skilled mobile phone repair technicians has grown exponentially. This section serves as a foundational introduction to the world of mobile phone repair, shedding light on its scope, importance, challenges, and the opportunities it presents.

Understanding the Scope of Mobile Phone Repair

The Need for Mobile Phone Repair Technicians: In an era where nearly everyone owns a mobile phone, the demand for repair services is constant. Mobile phone repair technicians play a vital role in addressing issues ranging from hardware malfunctions to software glitches. They are the troubleshooters who breathe new life into malfunctioning devices, ensuring that people stay connected and productive.

The Ubiquity of Mobile Phones: Mobile phones have transcended their original purpose as communication devices. They are now personal assistants, cameras, entertainment centers, and more. Their widespread use means that virtually every community requires mobile phone repair services, making it a profession with global relevance.

The Economic Opportunities in Mobile Repair: Mobile phone repair isn't just a service; it's also a thriving industry. Entrepreneurs can establish profitable repair businesses, and skilled technicians are in high demand. For individuals looking for a rewarding career or an entrepreneurial venture, mobile phone repair offers substantial economic opportunities.

Limitations and Challenges

Not All Problems Can Be Fixed: It's important to recognize that not every mobile phone issue can be resolved. Some problems may be irreparable due to hardware limitations or extensive damage. Understanding these limitations is crucial for managing customer expectations and offering ethical services.

Ethical Considerations in Mobile Phone Repair: Repair technicians must operate with integrity. This includes respecting customer privacy and ensuring data security. Ethical conduct not only builds trust but also safeguards against potential legal issues.

Legal Regulations and Licensing: Mobile phone repair may be subject to regional regulations, licensing requirements, and environmental standards. Technicians should be aware of and comply with relevant laws and regulations in their area of operation.

Importance of Mobile Phone Repair Technicians

Bridging the Digital Divide: Mobile phone repair technicians contribute to bridging the digital divide by ensuring that individuals from all walks of life have access to functioning devices. This accessibility is essential for education, job opportunities, and staying connected with loved ones.

Contributing to Environmental Sustainability: Repairing and extending the lifespan of mobile phones reduces electronic waste. Mobile phone repair, when done responsibly, aligns with sustainability goals by minimizing the disposal of electronic components.

Building a Satisfying Career: For those passionate about technology and problem-solving, mobile phone repair can be a fulfilling and financially rewarding career path. The constant evolution of mobile technology keeps the profession dynamic and engaging.

Course Overview

The Structure of this Course Manual: This course manual is designed to provide aspiring mobile phone repair technicians with a comprehensive guide. It is organized into sections that cover essential topics, from the basics of mobile phones to advanced repair techniques.

Learning Objectives and Outcomes: Throughout this course, you will have clear learning objectives and expected outcomes for each section. These objectives will guide your learning and help you assess your progress.

How to Get the Most Out of this Manual: To make the most of this course, active engagement and hands-on practice are key. Take notes, ask questions, and practice the techniques and skills outlined in this manual. Your commitment to learning will determine tyour success in mobile phone repair.

As you embark on your journey to become a skilled mobile phone repair technician, remember that this manual is your roadmap. It will equip you with the knowledge and skills needed to excel in this rewarding field. With dedication, practice, and a commitment to safety and ethics, you can become a proficient mobile phone repair technician ready to tackle the challenges and opportunities in this dynamic industry.

Mobile Phones: An Overview

Mobile phones, often referred to simply as "cell phones" in some regions, have transformed the way we communicate, work, and interact with the world. In this section, we delve into the multifaceted world of mobile phones, providing a comprehensive overview that covers their definition, purpose,

standards, evolution, and the necessity they represent in modern life.

Definition and Purpose

Definition: A mobile phone, in its most fundamental form, is a portable electronic device designed primarily for voice communication over long distances. However, modern mobile phones have evolved far beyond this initial purpose. They are now versatile gadgets that combine a wide range of functions, including texting, web browsing, photography, navigation, and much more.

Purpose: Mobile phones serve as lifelines in today's fast-paced world. They enable individuals to stay connected with family, friends, and colleagues regardless of geographical barriers. They facilitate instant access to information, entertainment, and services, making them indispensable tools for both personal and professional use.

Standards and Regulations

Standards: The mobile phone industry adheres to a set of technical standards that dictate the design, functionality, and compatibility of mobile devices. These standards ensure that mobile phones produced by different manufacturers can communicate with each other seamlessly. Common standards

include GSM (Global System for Mobile Communications) and CDMA (Code Division Multiple Access).

Regulations: Mobile phones are subject to various regulations, including spectrum allocation, wireless communication standards, and consumer protection laws. Regulatory bodies such as the Federal Communications Commission (FCC) in the United States play a crucial role in ensuring the safety, functionality, and fairness of the mobile phone ecosystem.

Evolution of Mobile Phone Technology

The journey of mobile phone technology has been nothing short of extraordinary. From the bulky, analog devices of the past to the sleek, powerful smartphones of today, mobile phones have undergone a remarkable transformation. Key milestones in this evolution include:

- **First-Generation (1G) Phones**: Introduced in the 1980s, 1G phones allowed voice communication over cellular networks but lacked the data capabilities of modern devices.

- **Second-Generation (2G) Phones:** The advent of 2G brought digital communication, improved call quality, and basic text messaging.

- **Third-Generation (3G) Phones:** 3G technology introduced faster data transfer speeds, enabling mobile internet access and multimedia messaging.

- **Fourth-Generation (4G) Phones:** 4G marked a significant leap, offering high-speed internet, video streaming, and app downloads.

- **Fifth-Generation (5G) Phones:** 5G technology promises even faster data rates, low latency, and support for emerging technologies like augmented reality (AR) and the Internet of Things (IoT).

Necessity in Modern Life

Mobile phones have become an essential part of contemporary life for several reasons:

- **Communication:** They provide instant and reliable communication, bridging geographical distances and fostering connections.

- **Information Access:** Mobile phones offer access to vast amounts of information, enhancing education, work, and personal knowledge.

- **Productivity:** They are powerful tools for productivity, enabling remote work, scheduling, and organization.

- **Entertainment:** Mobile phones serve as entertainment hubs, with capabilities for gaming, video streaming, music playback, and photography.

- **Navigation:** GPS technology in mobile phones has revolutionized navigation, making it easier to find locations and get directions.

In summary, mobile phones are no longer mere communication devices; they are indispensable companions that empower us to navigate the complexities of modern life. Their evolution continues, promising even more transformative capabilities in the future. Understanding these devices is the first step in unlocking their full potential and becoming proficient in mobile phone repair.

Safety Precautions in Mobile Phone Repair

Mobile phone repair involves working with delicate electronic components and potentially hazardous materials. Safety precautions are paramount to protect both the technician and the device being repaired. In this section, we explore the critical safety measures that should be followed when performing mobile phone repairs.

Handling Hazardous Materials

1. **Electrostatic Discharge (ESD) Protection:** Static electricity can severely damage sensitive electronic components. Technicians should use ESD-safe tools, wear anti-static wrist straps, and work on ESD-safe surfaces to prevent static discharge.

2. **Battery Safety:** Mobile phone batteries can be volatile if mishandled. Avoid puncturing, crushing, or exposing them to extreme temperatures. Always use the correct tools for battery removal and replacement.

3. **Chemical Hazards:** Some cleaning solvents and adhesives used in repair can be harmful if inhaled or absorbed through the skin. Work in well-ventilated areas and use personal protective equipment (PPE) like gloves and masks when necessary.

Safe Work Environment

4. **Clean and Organized Workspace:** A clutter-free and organized workspace reduces the risk of accidents. Ensure that tools and components are neatly arranged to prevent accidental spills or damage.

5. **Proper Lighting:** Adequate lighting is essential for precision work. Insufficient light can strain your eyes and lead to mistakes. Use task lighting as needed.

6. **Fire Safety:** Be cautious when using soldering equipment. Keep fire extinguishing equipment on hand, and know how to use it in case of a fire.

7. **First Aid Kit:** Always have a well-stocked first aid kit nearby in case of minor injuries. Familiarize yourself with basic first aid procedures.

Tools and Equipment Safety

8. **Tool Safety:** Use the right tools for the job. Improvising with tools that are not designed for mobile phone repair can lead to accidents and damage to the device.

9. **Unplug Devices:** Before working on a mobile phone, ensure it is powered off and disconnected from any power sources to prevent electrical shocks.

10. **Sharp Objects:** Be cautious with sharp tools like razor blades and screwdrivers. Keep them securely stored when not in use and handle them with care.

Personal Safety

11. **Protective Gear:** Depending on the task, wear appropriate safety gear, such as safety glasses, gloves, and a lab coat to protect against chemical splashes.

12. **Hygiene:** Wash your hands thoroughly after handling devices, especially if they have visible contaminants or adhesives. Avoid eating or drinking in the workspace.

13. **Training and Knowledge:** Ensure that you are properly trained and have the necessary knowledge and skills to perform the repair. Inexperienced attempts can lead to accidents and damage.

Emergency Preparedness

14. **Emergency Response Plan:** Have a clear plan in place for responding to emergencies, including accidents, fires, or unexpected device malfunctions.

15. **Emergency Contacts:** Keep contact information for emergency services and relevant professional contacts readily accessible.

By prioritizing safety precautions, mobile phone repair technicians can protect themselves, their workspace, and the devices they work on. Safety should always come first to ensure a successful and secure repair process.

Dismantling a Mobile Phone

Dismantling a mobile phone is a fundamental skill for any mobile phone repair technician. Whether you need to replace a faulty component or diagnose an issue, knowing how to safely take apart a device is crucial. In this section, we'll explore the steps and considerations involved in dismantling a mobile phone.

Tools and Equipment Needed

Before you begin dismantling a mobile phone, ensure you have the necessary tools and equipment, which may include:

1. **Screwdrivers:** Different types and sizes of screwdrivers are often required to remove screws of various shapes and sizes.

2. **Plastic Pry Tools:** These non-metallic tools are essential for safely opening the phone's casing without damaging it or internal components.

3. **Spudgers:** Spudgers help in prying apart components and connectors without causing damage.

4. **Tweezers:** Tweezers with fine tips are useful for handling small screws and delicate components.

5. **Anti-static Wrist Strap:** Prevent electrostatic discharge (ESD) by wearing an anti-static wrist strap connected to a grounded point.

6. **Magnifying Glass or Microscope:** These tools can assist in working with tiny components and connectors.

7. **Heat Gun or Heat Pad:** When dealing with glued components, gentle heating can make it easier to open the device.

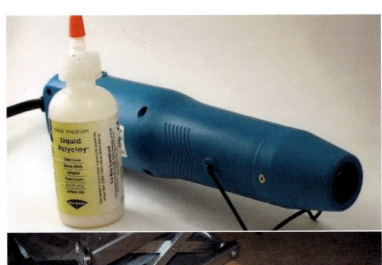

General Dismantling Procedure

Here's a general step-by-step guide for dismantling a mobile phone:

1. Power Off Ensure the phone is powered off and disconnected from any power sources.

2. **Remove the Back Cover:** In many cases, the back cover can be removed by prying it gently with plastic pry tools. Be cautious around any clips or latches.

3. **Remove Screws:** Use the appropriate screwdriver to carefully remove any screws securing the phone's casing. Keep track of screw locations and their sizes, as they may vary.

4. **Separate the Casing:** Once the screws are removed, gently separate the phone's casing. Be mindful of any cables or connectors that may still be attached.

5. **Disconnect Battery:** If possible, disconnect the battery to prevent electrical mishaps during the dismantling process.

6. **Access Internal Components:** With the casing removed, you can access the internal components of the phone. These may include the motherboard, battery, camera modules, and more.

7. **Handle Components Carefully:** When handling internal components, use anti-static precautions, and be gentle to avoid damage.

8. **Document and Label:** It's a good practice to document your progress and label connectors or components that you disconnect. This makes reassembly easier.

9. **Replace or Repair:** Once you've identified the issue or component that needs attention, proceed with the necessary repair or replacement.

Variations in Dismantling Methods

It's important to note that not all mobile phones are constructed the same way. The dismantling process can vary significantly between different models and manufacturers. Therefore, always refer to specific device documentation or guides when available. Manufacturers may have their own unique methods and components.

In summary, dismantling a mobile phone is a critical skill in the world of mobile phone repair. It allows you to access internal components, diagnose issues, and perform necessary repairs or replacements. Always prioritize safety, organization, and precision during the dismantling process to ensure a successful repair and minimize the risk of damage to the device.

Reassembling a Mobile Phone

Reassembling a mobile phone is the crucial final step in the repair process. It requires precision, organization, and attention to detail to ensure that all components are correctly placed and secured. In this section, we'll guide you through the steps and best practices for reassembling a mobile phone.

Preparation and Tools

Before you begin reassembling the phone, gather the following tools and materials:

1. **Components:** Ensure that all replacement components are readily available and in good condition.

2. **Screws:** Organize and label screws according to their sizes and locations. It's crucial not to mix them up.

3. **Plastic Pry Tools:** These tools help snap the casing back together without causing damage

4. **Tweezers:** Fine-tipped tweezers are handy for precise placement of small components.

5. **Anti-static Wrist Strap:** Wear an anti-static wrist strap connected to a grounded point to prevent electrostatic discharge (ESD).

6. Workspace: Work in a clean, well-lit area with a clutter-free workspace.

Reassembly Steps

1. **Component Placement:** Begin by carefully placing the main components back into the phone, starting with the motherboard. Ensure that connectors align properly with their respective slots.

2. **Cable Connections:** Reconnect any cables or ribbons that were detached during the dismantling process. Use caution and make sure they are securely fastened.

3. **Battery Connection:** If the battery was disconnected, reconnect it. Always do this after connecting other components to avoid potential electrical issues.

4. **Screw Installation:** Use your labeled screws to reattach components to the motherboard and casing. Follow the order specified in the disassembly guide or your notes. Tighten screws firmly but avoid over-tightening, which can damage components or strip threads.

5. **Clips and Snaps:** Pay attention to any clips or snaps that secure the casing. Ensure that they engage correctly to hold the phone's parts together.

6. **Casing Alignment:** Carefully align the phone's casing and gently snap it back into place. Apply even pressure to avoid any stress on the components.

7. **Final Checks:** Before powering on the phone, perform a thorough visual inspection. Make sure there are no loose components, cables, or screws left behind.

8. **Testing:** Power on the phone and test all functions, including the touchscreen, buttons, cameras, speakers, and connectivity. Verify that the device operates correctly.

Best Practices for Reassembly

- **Patience:** Take your time during the reassembly process. Rushing can lead to mistakes and potential damage.

- **Double-Check Connections:** Ensure that all cables and connectors are correctly seated. Loose connections can cause malfunctions.

- **Follow Documentation:** Refer to any repair guides, diagrams, or documentation provided by the manufacturer or your own notes.

- **Cleanliness:** Keep components and your workspace clean to prevent dust or debris from getting inside the phone.

- **Test Before Final Closure:** Always test the device before fully closing the casing. This allows you to address any issues without having to disassemble the device again.

Reassembling a mobile phone is a delicate and precise task. When done correctly, it restores the device to its functional

state, and the repair process is complete. Remember to maintain a methodical and organized approach to ensure a successful reassembly, and always prioritize safety and accuracy.

Mobile Phone Components

Mobile phones are complex devices composed of various components that work together to provide a wide range of functions. Understanding these components is essential for mobile phone repair technicians. In this section, we will delve into the key mobile phone components, their functions, and why they are crucial to the device's operation.

1. Battery:

   - Function: Provides power to the phone.

   - Importance: Without a functional battery, the phone cannot operate. Battery health affects overall device performance and usage time.

2. Motherboard (Mainboard):

   - Function: The central circuit board that houses the CPU, RAM, and other essential components.

   - Importance: The motherboard is the heart of the phone, responsible for processing data and controlling all functions.

3. CPU (Central Processing Unit):

   - Function: Executes instructions and processes data.

   - Importance: Determines the phone's processing speed and overall performance.

4. RAM (Random Access Memory):

   - Function: Provides temporary storage for running applications and data.

   - Importance: Affects multitasking capabilities and responsiveness.

5. Storage (e.g., NAND Flash):

   - Function: Stores the phone's operating system, apps, and user data.

   - Importance: Determines the device's storage capacity and data access speed.

6. Display (LCD or OLED):

   - Function: Provides visual output and touch interface.

   - Importance: Essential for user interaction and displaying information.

7. Touchscreen Digitizer:

   - Function: Detects touch input and translates it into digital signals.

   - Importance: Enables touch-based interaction with the phone.

8. Cameras (Rear and Front):

   - Function: Capture photos and videos.

   - Importance: Vital for photography, video calls, and various apps.

9. Microphone and Speaker:

   - Function: Record and play back audio.

   - Importance: Facilitates voice calls, video recording, and audio playback.

10. Buttons and Sensors (e.g., Power, Volume, Proximity):

   - Function: Enable user control and gather data from the phone's environment.

   - Importance: Essential for navigation, power control, and sensor-based features.

11. SIM Card and Memory Card Slots:

   - Function: Accept SIM cards for network connectivity and memory cards for additional storage.

   - Importance: Enable cellular communication and expand storage capacity.

12. Antennas:

- Function: Transmit and receive wireless signals (cellular, Wi-Fi, Bluetooth, GPS).

- Importance: Crucial for connectivity and location-based services.

13. Vibrator Motor:

- Function: Generates vibrations for notifications and haptic feedback.

- Importance: Enhances user experience and provides alerts.

4. Housing and Casing:
- Function: Enclose and protect internal components.

- Importance: Provides physical durability and aesthetics.

15. Connectors and Ports (e.g., Charging Port, Headphone Jack)

- Function: Enable charging, data transfer, and peripheral connections.

- Importance: Facilitate device charging and external device connections.

16. Adhesive and Thermal Materials:

- **Function:** Secure components and manage heat.

- **Importance:** Ensure component stability and prevent overheating.

17. PCB (Printed Circuit Board):

   - Function: Connects and facilitates communication between various components.

   - Importance: Forms the electrical backbone of the device.

Understanding these mobile phone components and their roles is essential for diagnosing and repairing issues. Mobile phone repair technicians must be proficient in identifying, testing, and replacing these components to ensure the proper functioning of the device. Additionally, staying up-to-date with advancements in mobile technology is crucial as new components and features are continually introduced in the ever-evolving world of mobile phones.

Battery Details and Compatibility

Understanding battery voltage, compatibility, and terminal configurations is crucial in mobile phone repair. Let's delve into these details:

Battery Voltage:

- Mobile phone batteries typically have a nominal voltage of 3.7V. This voltage is sufficient to power the phone effectively. A battery below 3.6V may still power the phone but might show a low battery warning. Batteries in the 3.6V to 3.65V range can

usually power the phone but may not provide optimal performance.

- Batteries with a voltage above 3.7V are considered over-voltage and can lead to issues such as excessive heat in the motherboard. It's crucial to use batteries with the correct voltage rating for the phone.

Battery Model and Compatibility:

- Each phone model has a specific battery model number and often mentions the manufacturer's name. For example, a Tecno battery with the model BL-34H and an Itel battery with the same model BL-34H are compatible with each other and can be used interchangeably.

- However, if two batteries have the same manufacturer name but different model numbers, they may not be compatible with each other. Always use the recommended battery model for a specific phone.

Terminal Configuration:

- Batteries may have two, three, or four terminals.

- Two-terminal batteries have a positive and a negative terminal.

- Three-terminal batteries have a negative, positive, and two indicators: Battery State Indicator (BSI) and Battery Temperature Indicator (BTI), often combined in the middle positive terminal but separated in four-terminal batteries.

- Four-terminal batteries typically have a negative and positive terminal and additional terminals for BSI and BTI functions.

Test Points for Battery:

- Mobile phones have test points labeled according to battery specifications. Positive is often labeled as "VBAT" and Negative is labeled as "GND" (Ground). These test points are usually larger and easier to identify compared to others.

- Test points may be connected to resistors or labeled accordingly to guide technicians in connecting the battery correctly.

Understanding these battery details is crucial for selecting the right battery for replacement and ensuring safe and efficient operation of the mobile phone. Additionally, following manufacturer recommendations and specifications is essential to prevent compatibility issues and potential damage to the device.

Using a Multimeter for Fault Finding in Mobile Phone Repair

A multimeter is an indispensable tool for mobile phone repair technicians. It allows you to measure various electrical parameters and diagnose faults accurately. In this section, we'll explore how to use a multimeter effectively for fault finding in mobile phone repair.

1. Safety First:

- Always ensure the phone is powered off and disconnected from any power source before using a multimeter.

- Wear an anti-static wrist strap to prevent electrostatic discharge (ESD) that could damage sensitive components.

2. Select the Correct Multimeter Settings:

- Set the multimeter to the appropriate mode for your measurement. Common modes include voltage (V), current (A), resistance (Ω), and continuity (a sound or visual indicator for a closed circuit).

3. Voltage Measurement:

- To measure voltage, place the multimeter probes on the points you want to test. For example, to check battery voltage, place one probe on the positive terminal and the other on the negative terminal.

- Ensure that the voltage falls within the expected range for the component. Abnormal voltage readings can indicate issues with power supply or components.

4. Resistance Measurement:

- To measure resistance, disconnect the power source and ensure the component is not powered.

- Place the multimeter probes at either end of the component or circuit you want to test.

- Compare the resistance reading to the expected value. A reading significantly different from the expected value may indicate a faulty component.

5. Continuity Testing:

- Continuity testing helps you determine if a circuit is open or closed. It's particularly useful for identifying broken connections or damaged wires.

- Set the multimeter to the continuity mode (often represented by a sound or a diode symbol).

- Touch the probes to opposite ends of the circuit or component. If the circuit is closed (continuous), you will hear a beep or see a visual indicator.

6. Current Measurement:

- Measuring current often requires breaking the circuit to insert the multimeter in series. This is more common in advanced troubleshooting.

- Be cautious when measuring current, as connecting the multimeter improperly can damage the device.

7. Record and Document:

- Keep a record of your measurements and findings. This documentation can help you track the diagnostic process and pinpoint the fault more effectively.

8. Systematic Approach:

- Approach fault finding systematically, starting with the most common issues and gradually moving to more complex ones. Begin with power-related problems and work your way through various components.

9. Be Patient and Thorough:

- Fault finding can be time-consuming and may require multiple measurements and tests. Be patient and thorough to ensure you identify and resolve the issue accurately.

10. Follow Safety Guidelines:

- Always adhere to safety guidelines when working with a multimeter and electrical components. Avoid short circuits, and be cautious around live circuits to prevent injury.

By using a multimeter effectively, mobile phone repair technicians can pinpoint faults, diagnose issues, and make precise repairs. This tool is an essential part of the diagnostic process, helping ensure the proper functioning of mobile devices.

Common Mobile Phone Faults and Solutions

Mobile phones are complex devices, and like all electronics, they can experience a range of issues. Mobile phone repair technicians often encounter these common faults and must have the knowledge and skills to diagnose and resolve them. Here's a list of some prevalent mobile phone faults and their potential solutions:

1. Battery Drainage:

   - Fault: Rapid battery drainage, even when the phone is not in use.

   - Solution:

     - Check for power-hungry apps running in the background and close them.

     - Adjust screen brightness and screen timeout settings.

     - Replace the battery if it's old and no longer holding a charge.

2. Screen Issues:

   - Fault: Cracked or unresponsive touchscreen, dead pixels, or display discoloration.

   - Solution:

     - Replace the touchscreen or LCD display if physically damaged.   - Perform a software reset or update to address unresponsive touch issues.

- Check and reseat display connectors.

3. Charging Problems:

  - Fault: Phone not charging or slow charging.

  - Solution:

  - Inspect the charging port for debris or damage and clean or replace it if necessary.

  - Use a different charging cable and adapter to rule out faulty accessories.

  - Replace the battery if it's not holding a charge.

4. Overheating:

  - Fault: The phone gets excessively hot during normal use.

  - Solution:

  - Close resource-intensive apps running in the background.

  - Ensure proper ventilation and avoid using the phone on soft surfaces that can block airflow.

  - Check for a malfunctioning battery or internal components and replace if necessary.

5. No Signal or Network Issues:

  - Fault: Loss of network signal or constant network drops.

- Solution:

   - Restart the phone and try toggling airplane mode.

   - Check for network outages in your area.

   - If the issue persists, contact your service provider or check for a faulty SIM card.

6. Audio Problems:

   - Fault: No sound during calls, distorted audio, or no audio output from speakers.

   - Solution:

   - Check for debris or blockage in the earpiece, microphone, or speaker openings and clean if necessary.

   - Test with headphones to determine if the issue is with the internal speakers.

   - Update or reinstall audio drivers or perform a factory reset as a last resort.

7. Slow Performance:

   - Fault: The phone operates slowly, freezes, or experiences lag.

   - Solution:

   - Clear cache and unnecessary data.

   - Uninstall resource-intensive or unused apps.

- Factory reset the phone to restore its performance, but back up data first.

8. Camera Problems

- Fault: Blurry images, camera app crashes, or issues with focusing.

- Solution:

  - Clean the camera lens.

  - Check for software updates for camera-related issues.

  - Replace the camera module if hardware-related problems persist.

9. Software Glitches:

- Fault: Frequent app crashes, freezing, or the phone restarting unexpectedly.

- Solution:

  - Update the operating system and apps to the latest versions.

  - Clear app cache and data.

  - If issues persist, consider a factory reset, but backup data first.

It's important to note that these solutions provide general guidance for common mobile phone faults. The specific solution may vary based on the phone model and the nature of the problem. Mobile phone repair technicians should have the skills to diagnose issues accurately and apply appropriate solutions to ensure the device's proper functionality.

Water Damage and Solutions for Mobile Phones

Water damage is a common and potentially severe issue for mobile phones. Accidents happen, and when phones come into contact with water, quick action is crucial to minimize damage. Here's how to handle water damage and potential solutions:

1. Immediate Actions:

  - Fault: Phone exposed to water, moisture, or any liquid.

  - Solution:

    - Power off the phone immediately, if it's still on.

    - Remove the phone from water or liquid as quickly as possible.

    - Remove any peripherals, such as the SIM card and memory card.

2. Do Not Turn It On:

  - Fault: Attempting to power on the phone after water exposure.

  - Solution:

    - Do not try to turn on the phone or press any buttons. This can cause short circuits and further damage.

3. Remove the Battery (If Removable):

- Fault: Leaving the battery connected after water exposure.

- Solution:

- If the phone has a removable battery, take it out to prevent electrical damage.

4. Disassemble (If Possible):

- Fault: Keeping the phone fully assembled with trapped moisture.

- Solution:

- If you have experience or a repair technician nearby, disassemble the phone to allow proper drying.

5. Dry the Phone:

- Fault: Allowing the phone to remain wet.

- Solution:

- Gently blot the phone with a dry cloth or paper towel to remove surface moisture.

- Avoid using heat sources like hairdryers or ovens, as excessive heat can damage internal components.

6. Rice or Silica Gel:

- Fault: Not using desiccants to absorb moisture.

- Solution:

   - Place the phone in a bag of uncooked rice or silica gel packets. These desiccants can help absorb remaining moisture.

   - Leave the phone in the bag for at least 24-48 hours.

7. Professional Repair:

   - Fault: Attempting DIY repairs without expertise.

   - Solution:

   - If the phone does not power on or function properly after drying, seek professional repair assistance.

   - Technicians can disassemble the device, clean internal components, and assess damage for repair or replacement.

8. Data Backup:

   - Fault: Not backing up data before attempting to dry or repair the phone.

   - Solution:

   - If the phone is non-responsive, it may be necessary to recover data through professional data recovery services.

9. Prevention:

   - Fault: Not taking precautions to prevent water damage.

   - Solution:

- Invest in a waterproof phone case.

- Be cautious around water sources, and avoid using your phone near pools, sinks, or in heavy rain.

10. Consider Waterproof Phones:

   - Fault: Frequent water exposure.

   - Solution:

   - If you're in an environment where water exposure is common, consider purchasing a waterproof or water-resistant phone.

Remember that the severity of water damage can vary, and not all phones can be saved. Quick and appropriate actions can significantly increase the chances of successful recovery. However, it's always best to consult with a professional technician for water damage assessment and repair if you're unsure about handling it yourself.

Professional Solution for Water-Damaged Mobile Phones (Technician's Guide)

Repairing a water-damaged mobile phone as a technician requires a systematic and careful approach. Here's a step-by-step method to professionally address water damage:

1. Safety Precautions:

- Before you begin, ensure safety by wearing an anti-static wrist strap to prevent electrostatic discharge (ESD) damage.

2. Isolate the Phone:

- Remove the phone from any power source or connection.

- Power off the phone if it's still on.

3. Remove the Battery (If Possible):

- If the phone has a removable battery, take it out to prevent electrical damage.

4. Disassemble the Phone:

- If you have experience and the necessary tools, carefully disassemble the phone. Refer to manufacturer guides or diagrams if available.

- Take pictures of the disassembly process to aid in reassembly later.

5. Dry the Phone:

- Use compressed air or a can of electronic contact cleaner to blow out any excess water or moisture.

- Gently blot the phone with a lint-free, absorbent cloth or paper towel to remove surface moisture.

6. Remove Components:

- Remove any detachable components, such as the SIM card, memory card, and peripherals.

- Disconnect and remove any components with visible moisture, such as the battery, screen, and connectors.

7. Inspect for Corrosion and Damage:

- Carefully inspect the internal components for signs of corrosion or water damage.

- Clean corrosion using a soft brush and isopropyl alcohol (at least 90% concentration).

8. Submerge in Isopropyl Alcohol (Optional):

- In some cases, especially if there's significant water damage, you can submerge the disassembled components in isopropyl alcohol. This can help displace water and accelerate drying.

- Ensure the alcohol used is at least 90% concentration, and soak components for a short time (e.g., 5-10 minutes).

9. Thorough Drying:

- Allow all components to air-dry in a clean, dry, and well-ventilated area for at least 24-48 hours. Use desiccants like silica gel packets to aid in moisture removal.

- Ensure the phone is completely dry before reassembly.

10. Reassemble the Phone:

- Reconnect and reassemble all components carefully, following your documentation or reference pictures.

11. Test the Phone:

- Power on the phone and test all functions, including touchscreen, buttons, camera, and connectivity.

- Ensure that the phone is operating normally and that there are no issues with any components.

12. Data Recovery (If Needed):

- If the phone does not power on or function correctly after drying, consider professional data recovery services to retrieve data if required.

13. Customer Communication:

- Inform the phone owner of the process and results, especially if there's been significant damage or data loss.

14. Warranty Considerations:

- If the phone is under warranty, check the manufacturer's warranty policy regarding water damage before proceeding with repair.

Please note that water damage can vary in severity, and not all phones can be successfully repaired. Professional technicians should exercise caution and expertise throughout the repair process. If you're uncertain about any step or if the water damage is extensive, it's advisable to consult with a specialist or consider the possibility that the phone may be irreparable.

Advanced Troubleshooting Techniques in Mobile Phone Repair

Advanced troubleshooting techniques are essential for mobile phone repair technicians when dealing with complex or elusive issues. These techniques go beyond basic diagnostics and require a deeper understanding of mobile phone components and software. Here's a guide to advanced troubleshooting techniques in mobile phone repair:

1. Understanding Schematics:

- Mobile phone schematics provide detailed diagrams of the phone's circuitry. Technicians should learn to read and interpret schematics to pinpoint faults accurately.

2. Logic Board Diagnosis:

- Diagnosing issues on the logic board (motherboard) requires advanced skills. Technicians must be proficient in using tools like multimeters and oscilloscopes to identify faulty components, including ICs, capacitors, resistors, and more.

3. Advanced Software Diagnosis:

- Advanced technicians can analyze phone software at a deeper level. They use debugging tools and firmware analysis to identify and resolve software-related issues, including boot loops, crashes, and system errors.

4. JTAG and Chip-off Methods:

- In cases of severe software or hardware issues, technicians may use JTAG (Joint Test Action Group) or chip-off methods to access and recover data or repair the phone's firmware. These methods require specialized tools and expertise.

5. Data Recovery:

- Advanced technicians can recover data from damaged or non-functional phones using techniques such as chip-off data recovery, NAND flash memory chip reading, or specialized software tools.

6. Component-Level Repairs:

- Skilled technicians can perform component-level repairs, such as reballing or replacing individual ICs or damaged connectors on the logic board. This is a highly specialized skill.

7. EMI Shielding and RF Troubleshooting:

- Advanced troubleshooting includes diagnosing and resolving electromagnetic interference (EMI) issues and radio frequency (RF) signal problems, which can affect network connectivity.

8. Deep Firmware Modifications:

- Some repairs may involve deep firmware modifications, such as flashing custom ROMs, unlocking bootloaders, or patching firmware to bypass security locks. These techniques require advanced software skills.

9. Board-Level Soldering and Microsoldering:

- Skilled technicians may perform microsoldering to replace tiny components on the logic board. This is a delicate process and typically reserved for repairing intricate hardware faults.

10. Signal Tracing and Oscilloscope Analysis:

- Advanced technicians use oscilloscopes to trace signals through the phone's circuitry, helping identify where a signal is lost or compromised.

11. Advanced Tool Use:

- Mastery of specialized tools like ISP (In-System Programming) adapters, eMMC (embedded

- MultiMediaCard) programmers, and BGA (Ball Grid Array) rework stations is essential for advanced troubleshooting.

12. Firmware Reverse Engineering:

- In certain cases, technicians may need to reverse engineer firmware to understand its workings better and make modifications for specific repairs or customizations.

13. Advanced Water Damage Recovery:

- Beyond basic water damage recovery, advanced technicians may use ultrasonic baths and more sophisticated techniques to clean and repair water-damaged components.

14. Advanced Microscope Use:

- Precision microscope examination is crucial for identifying microscopic faults and damage on small components and solder joints.

Advanced troubleshooting techniques require a high level of expertise, practice, and access to specialized equipment and software tools. Technicians should continually update their skills and knowledge to stay current with evolving mobile phone technologies. These advanced skills are invaluable for addressing complex and challenging mobile phone repair scenarios.

Additional Resources for Mobile Phone Repair Technicians

Mobile phone repair is a dynamic field that continually evolves with new technologies and techniques. To stay informed, enhance skills, and troubleshoot effectively, technicians can access various resources. Here's a list of valuable additional resources:

1. Repair Manuals and Guides:

- Manufacturer repair manuals and official documentation provide in-depth information on repairing specific phone models. These manuals often include schematics and detailed procedures.

2. Online Forums and Communities:

- Websites and forums like XDA Developers, iFixit, and specialized mobile repair communities offer discussions, guides, and solutions to common and advanced repair issues. Technicians can interact with peers and seek advice.

3. Repair Schools and Training Courses:

- Consider enrolling in mobile phone repair courses offered by technical schools or online platforms. These courses cover fundamental and advanced repair techniques.

4. YouTube Tutorials:

- YouTube hosts numerous video tutorials on mobile phone repair. Technicians can watch step-by-step guides for specific repairs and gain practical insights.

5. Books and E-books:

- Mobile phone repair books, both in print and digital formats, provide comprehensive information on repair techniques, component-level repairs, and troubleshooting.

6. Trade Publications:

- Subscribing to trade magazines and publications like "Mobile Repair" or "Phone Technician's Journal" keeps technicians updated on industry trends, news, and techniques.

7. Online Courses and Certifications:

- Platforms like Coursera, Udemy, and LinkedIn Learning offer online courses and certifications in mobile phone repair, electronics, and related subjects.

8. Manufacturer Websites:

- Official websites of mobile phone manufacturers often contain support sections with repair guides, software updates, and technical documentation.

9. Mobile Repair Tools and Equipment Suppliers:

- Suppliers like iFixit, Union Repair, and eTech Parts provide tools, components, and equipment tailored for mobile phone repair.

10. Repair Software and Diagnostic Tools:

- Software tools like Mobile Doctor, 3uTools, and UFI Box offer diagnostic and repair capabilities for various mobile phone issues.

11. Online Marketplaces:

- Online marketplaces such as eBay and Alibaba can be sources for replacement parts, components, and specialized tools.

12. Technical Blogs and Websites:

- Blogs and websites dedicated to mobile phone repair often publish articles, case studies, and tutorials covering advanced repair techniques.

13. Manufacturer Support and Training Programs:

- Some manufacturers offer training programs and resources for authorized repair technicians. Technicians can inquire about these programs to gain access to official repair materials.

14. Social Media Groups:

- Social media platforms like Facebook and LinkedIn host groups and communities focused on mobile phone repair. Technicians can join these groups to network and share knowledge.

15. Local Repair Associations:

- In some regions, repair associations or guilds offer resources, networking opportunities, and training programs for technicians.

16. Continuous Learning:

- Mobile phone technology is ever-evolving. Technicians should make an effort to stay updated with the latest advancements by attending workshops, webinars, and conferences.

Effective use of these additional resources can help mobile phone repair technicians expand their knowledge, troubleshoot complex issues, and provide high-quality repair services to customers. Staying informed and continuously learning is key to success in this dynamic field.

Conclusion

Mobile phone repair is a challenging and rewarding field that requires a combination of technical expertise, problem-solving skills, and continuous learning. This guide has provided an overview of essential topics for aspiring or experienced mobile phone repair technicians.

From understanding the basics of mobile phone components to advanced troubleshooting techniques, this manual has covered a wide range of topics, including safety precautions, dismantling and reassembling phones, and identifying common faults and solutions. Technicians must also be equipped with the knowledge to handle water damage and use multimeters effectively.

Moreover, the guide emphasized the importance of accessing additional resources, such as repair manuals, online forums, training courses, and specialized tools, to stay current in this ever-evolving field.

In conclusion, mobile phone repair is not only about fixing devices but also about helping individuals stay connected to their digital lives. By honing their skills, technicians can provide valuable services, whether it's rescuing data from a water-damaged phone, resolving complex software issues, or replacing faulty components.

To excel in mobile phone repair, continuous learning, practice, and adaptability are key. As technology continues to advance, the knowledge and expertise gained from this manual, combined with a commitment to professional growth, will enable technicians to thrive in this dynamic and essential industry.

About the Author

Felix Raluchukwu Okafor is a dedicated and passionate mobile phone repair technician hailing from Anambra state, Nigeria. His journey in the world of mobile phone repair began in the vibrant state of Rivers, Nigeria, where he received his training under the guidance of his esteemed mentor, Igboke Chiwendu Samuel, fondly known as Oga Sammy. It was in the year 2013 that Felix completed his apprenticeship at Sammy GSM Clinic.

With a thirst for knowledge and a desire to advance both professionally and academically, Felix ventured back to his hometown, Anambra State. Here, he pursued higher education at the Federal Polytechnic Oko, all while nurturing his growing passion for mobile phone repair.

In 2018, Felix embarked on an entrepreneurial journey by establishing his own mobile phone repair workshop in Awka, the capital city of Anambra State. His workshop, named JATO Communication, quickly became a trusted name in the field of mobile phone repair. Located in Dike Park, near the bustling Eke Awka market, JATO Communication has provided expert repair services to countless satisfied customers and continues to thrive under Felix's expert supervision.

Felix's dedication to his craft is evident in his unwavering love for his job. He approaches each repair with meticulous care and

a commitment to excellence, ensuring that every mobile device that passes through his workshop is restored to its optimal functionality.

Felix's vision extends beyond the present, with plans to take his mobile phone repair business to the next level in the future. His ambition and determination are driving forces that promise further success in his endeavors.

In conclusion, Felix Raluchukwu Okafor is not just a skilled technician but also an ambitious entrepreneur who stands as a testament to the potential of passion and hard work in achieving one's goals. His story is an inspiration to those who aspire to excel in the dynamic and ever-evolving field of mobile phone repair.

Printed in Poland
by Amazon Fulfillment
Poland Sp. z o.o., Wrocław